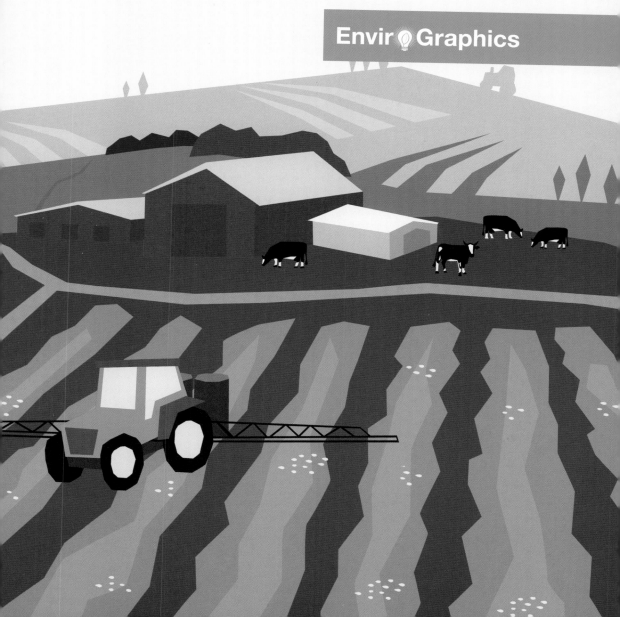

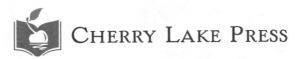

Published in the United States of America by Cherry Lake Publishing Group
Ann Arbor, Michigan
www.cherrylakepublishing.com

Reading Adviser: Marla Conn, MS, Ed., Literacy specialist, Read-Ability, Inc.
Photo Credits: ©grafikacesky/Pixabay, cover; ©OpenClipart-Vectors/Pixabay, cover; ©Shutterstock, cover; ©Shutterstock, 1; ©Clker-Free-Vector-Images/Pixabay, 5; ©gustavofer/Pixabay, 5; ©OpenClipart-Vectors/Pixabay, 5; ©PinkPanthress/Pixabay, 5; ©Shutterstock, 5; ©iStockphoto/Getty Images, 9; ©OpenClipart-Vectors/Pixabay, 9; ©iStockphoto/Getty Images, 11; ©grafikacesky/Pixabay, 11; ©OpenClipart-Vectors/Pixabay, 11; ©iStockphoto/Getty Images, 15; ©iStockphoto/Getty Images, 16; ©Shutterstock, 16; ©DigitalVision Vectors/Getty Images, 17; ©Clker-Free-Vector-Images/Pixabay, 18; ©Shutterstock, 18; ©Clker-Free-Vector-Images/Pixabay, 19; ©Janjf93/Pixabay, 19; ©mohamed_hassan/Pixabay, 19; ©Shutterstock, 19; ©Shutterstock, 21; ©Shutterstock, 23; ©Shutterstock, 27; ©Shutterstock, 28; ©Shutterstock, 29; ©DigitalVision/Getty Images, 30

Copyright ©2021 by Cherry Lake Publishing Group
All rights reserved. No part of this book may be reproduced or utilized in any form or by any means without written permission from the publisher.

Cherry Lake Press is an imprint of Cherry Lake Publishing Group.

Library of Congress Cataloging-in-Publication Data has been filed and is available at catalog.loc.gov

Cherry Lake Publishing Group would like to acknowledge the work of the Partnership for 21st Century Learning, a Network of Battelle for Kids. Please visit http://www.battelleforkids.org/networks/p21 for more information.

Printed in the United States of America
Corporate Graphics

TABLE OF CONTENTS

INTRODUCTION
What Is Agriculture?........................... 4

CHAPTER 1
A World of Agriculture...................... 6

CHAPTER 2
Issues with Agriculture.................... 14

CHAPTER 3
Possible Solutions............................26

ACTIVITY.. 30
LEARN MORE 31
GLOSSARY 32
INDEX... 32

INTRODUCTION

What Is Agriculture?

Agriculture means farming. Taking care of the soil and growing crops are part of agriculture. So is raising animals. Crops and animals are needed to feed people around the world. Agriculture also provides many jobs. It makes money for countries.

Some farming methods can be bad for the environment. Pollution and less space for wild plants and animals are top concerns. There are many things farmers can do to help with these issues. People can also make helpful choices in their everyday lives.

AGRICULTURAL PRODUCTS

Food from Plants

Fruits

Vegetables

Grains

Animal Products

Meat and fish

Dairy

Eggs

Materials

Cotton

Wool

Lumber

[ENVIRO-GRAPHICS]

A World of Agriculture

Organized agriculture began thousands of years ago. Over time, people have discovered new methods and technologies. Different countries focus on different products. It depends on the type of land and the weather of a place.

As the global population has increased, so has agriculture. The average U.S. farm has changed in many ways. In the 1800s, there were many more farms. Each one was fairly small. Over time, the number of farms has decreased, while each farm has grown in size.

TIMELINE

9000 BCE
The first evidence of farming is from 11,000 years ago.

5500 BCE
People begin to use **irrigation**. It leads to major growth in farming.

100s CE
With advanced tools and machines, ancient Romans begin to **industrialize** farming.

700s–1100s
People in the Middle East use science and technology to improve farming methods.

1500s–1600s
Many agricultural plants and animals are traded around the world. It is part of what is called the Columbian Exchange.

1700s–1800s
Farming technology and equipment become more advanced.

1970s
Scientists begin altering plants' **genes** directly. They create genetically modified organisms (GMOs).

Now
Large-scale farms in the United States are highly industrialized. Machines do much of the work.

2011, National Geographic; 2019, History Link 101; 2019, Encyclopædia Britannica

[ENVIRO-GRAPHICS]

COUNTRIES WITH THE MOST AGRICULTURAL LAND

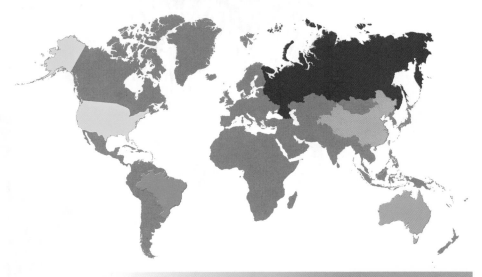

CHINA — 3,279,181 SQUARE MILES*

UNITED STATES — 2,521,913 SQUARE MILES

AUSTRALIA — 2,305,772 SQUARE MILES

BRAZIL — 1,761,873 SQUARE MILES

RUSSIA — 1,352,861 SQUARE MILES

*1 square mile = 2.6 square kilometers

2016, World Bank

MAJOR AGRICULTURAL PRODUCTS

United States: Corn
432,602,999 tons/year

United Kingdom: Milk
16,877,488 tons/year

Russia: Wheat
79,516,492 tons/year

Brazil: Sugar Cane
823,237,124 tons/year

Congo: Cassava
1,582,885 tons/year

China: Rice
283,678,337 tons/year

2018, Food and Agricultural Organization of the United Nations (FAO)

[ENVIRO-GRAPHICS]

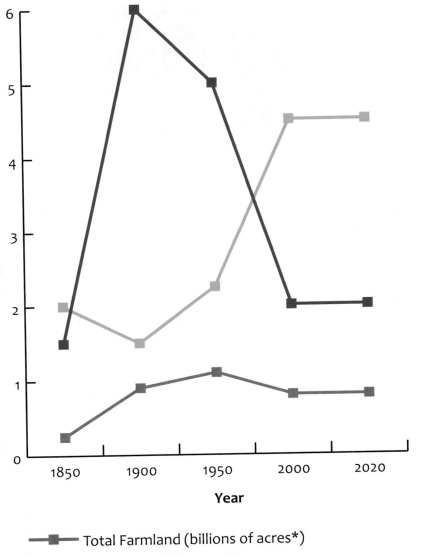

TODAY'S U.S. FARMS

One farm feeds about **166** people.

About **25%** of all farmers are new to farming.

The top **3** farm products are cattle, corn, and soybeans.

About **36%** of farmers are women.

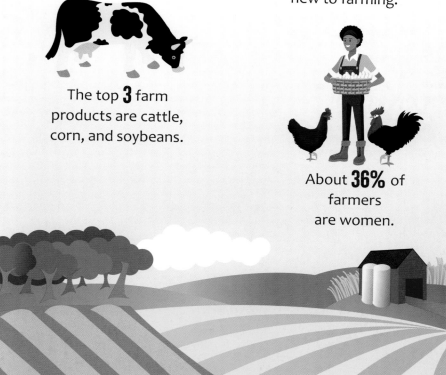

2017, U.S. Department of Agriculture

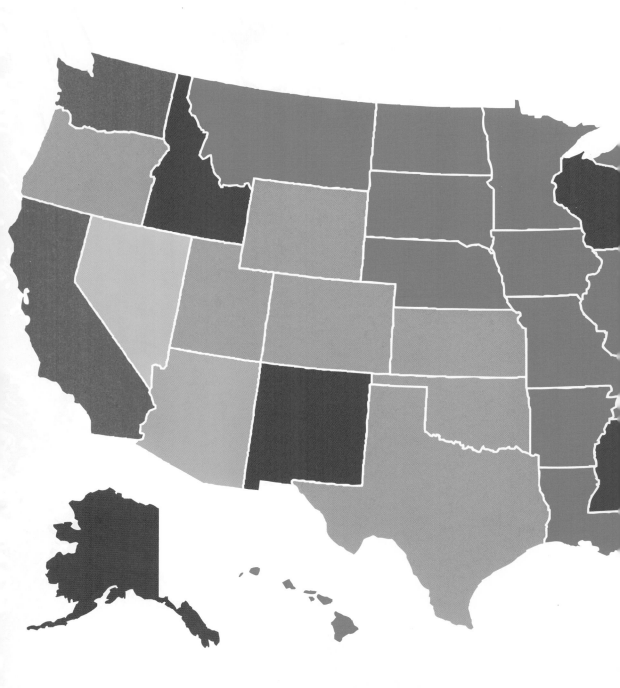

2019, Investment Watch

MOST VALUABLE AGRICULTURAL PRODUCTS BY STATE

 Fruit

 Cattle

 Other crops and hay

 Vegetables

 Fish

 Grains and beans

 Cow's milk

 Flowers and greenhouse plants

 Poultry and eggs

Issues with Agriculture

Agriculture is needed for people to live. But it comes with many issues. In some countries, food waste has become a major problem. Valuable water and land are used to produce the food. These resources are wasted too. As food waste increases, so does the need for more water and land. As more land is used for agriculture, it takes away land needed for wild plants and animals to thrive. Certain agricultural practices pollute Earth's water, land, and air.

Pesticides are sprayed onto crops to keep bugs away. Herbicides are used to kill weeds. These chemicals protect crops, but they can run off into the environment. They are not only toxic to the plants and animals they are trying to keep away. Many are also toxic to other animals and plants, as well as humans.

FAST FACTS

About **33%** of human food is lost or wasted every year.

Each person in Europe and North America wastes **209** to **254** pounds (95 to 115 kilograms) of food a year.

Every year, **$1 TRILLION** worth of food is wasted.

HOW FOOD IS WASTED

PRODUCTION
Unharvested crops
Damage by pests
Death from disease

GROCERY STORES
Overstocking
Expiration dates
Selection for ideal shape and size

PROCESSING
Bad storage conditions
Trimming
Overproduction

HOME
Oversized portions
Spoiled or ruined
Uneaten leftovers

2015, FAO

GLOBAL FOOD WASTE

Fruits and Vegetables
(equal to 3.7 trillion apples)

Cereals and Grains
(equal to 763 billion boxes of pasta)

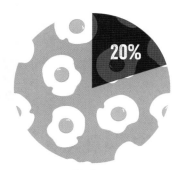

Dairy
(equal to 574 billion eggs)

Fish and Seafood
(equal to 3 billion Atlantic salmon)

Meat
(equal to 75 million cows)

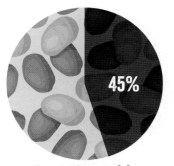

Root Vegetables
(equal to 1 billion bags of potatoes)

2012, Food and Agriculture Organization of the United Nations

LAND ISSUES

Habitat Loss

As of 2019, **1/2** of habitable land goes to agriculture.

A 2016 study found that more than **5,000 species** are threatened by agriculture.

Most habitat loss comes from livestock production. As of 2020, livestock use **77%** of the world's farming land.

Deforestation

In 2017, it was estimated that all rainforests would be cleared in **100 YEARS**.

As of 2013, about **80%** of **deforestation** in the Amazon was for cattle farming.

A 2015 study found that humans have cut down **46%** of the world's trees.

[ENVIRO-GRAPHICS]

WATER AND AIR ISSUES

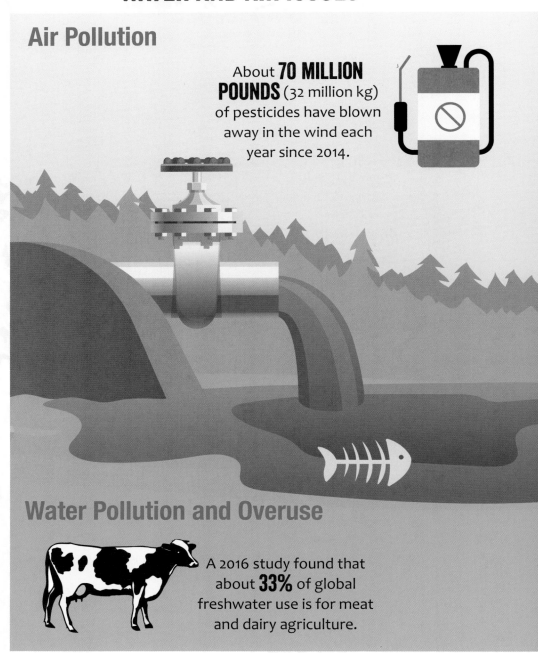

Air Pollution

About **70 MILLION POUNDS** (32 million kg) of pesticides have blown away in the wind each year since 2014.

Water Pollution and Overuse

A 2016 study found that about **33%** of global freshwater use is for meat and dairy agriculture.

 Agriculture is responsible for **16%** of one dangerous type of air pollution, said a 2009 study.

Since the early 2000s, U.S. food travels an average of **1,500 MILES** (2,414 kilometers) by truck, boat, and plane, which all put exhaust into the air.

Fertilizers and pesticides have made more than **400** ocean "dead zones," areas where normal sea life cannot exist.

As of 2019, meat and dairy cause **57%** of global water pollution.

[ENVIRO-GRAPHICS]

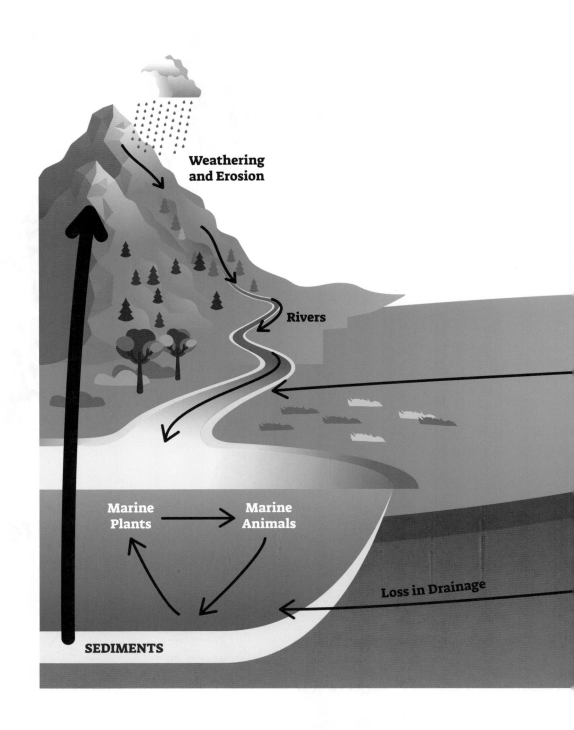

THE PESTICIDE CYCLE

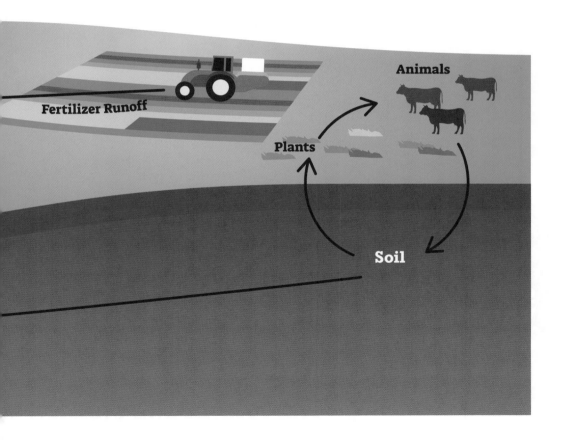

SYNTHETIC VS. NATURAL PESTICIDES

Synthetic
Made from chemicals
Work quickly
Cheap

Both
Can be toxic
Very effective

Natural
Made from natural ingredients
Work slowly
Expensive

TOTAL U.S. GAS EMISSIONS

Emissions are gases and particles that are put into the air. Emissions pollute the air.

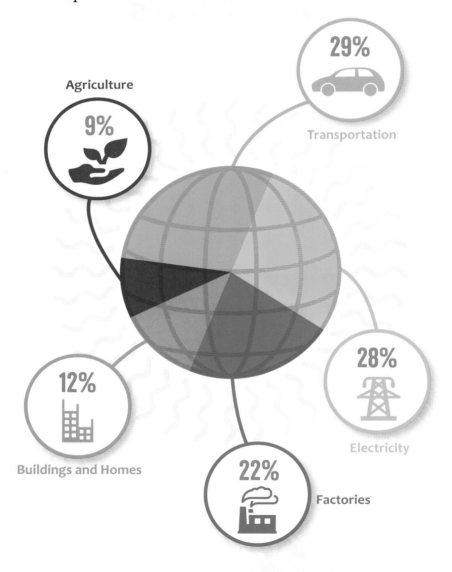

2019, U.S. Environmental Protection Agency

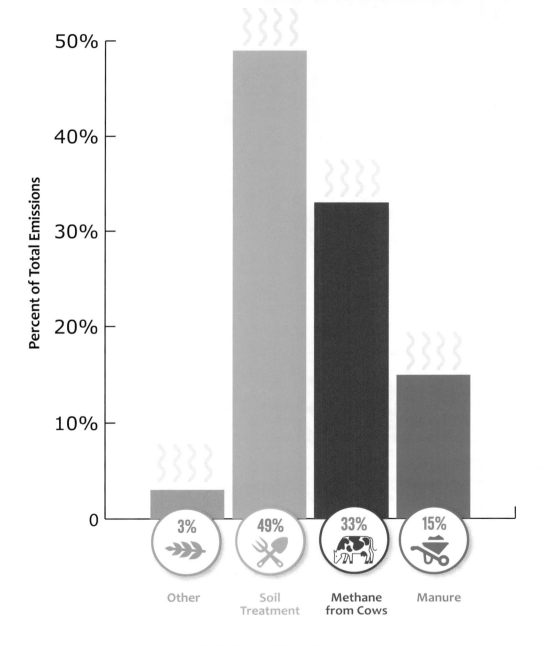

CHAPTER 3

Possible Solutions

As people learn more about agricultural issues, they come up with solutions. There are many things farmers and people can do to help.

Composting is one way to reduce food waste. Food scraps and certain leftovers can be put to good use. People can compost at home. Many **municipal** areas have composting programs too.

SOLUTIONS

What Farmers Can Do

- **Use fertilizers and pesticides carefully.** Timing can reduce the amount of chemicals needed.
- **Plant trees and shrubs around fields.** This can lead to 20% more **yield** in certain crops.
- **Feed livestock quality food.** Grazing cows produce 50% less methane.

What You Can Do

- **Reduce food waste.** Several U.S. agencies have a goal to reduce food waste to 109 pounds (49 kg) per person a year by 2030.
- **Eat less meat.** In 2019, beef alone was responsible for 3% of U.S. emissions.
- **Eat less processed food.** Food processing creates pollution.
- **Buy local.** This reduces food waste and pollution.

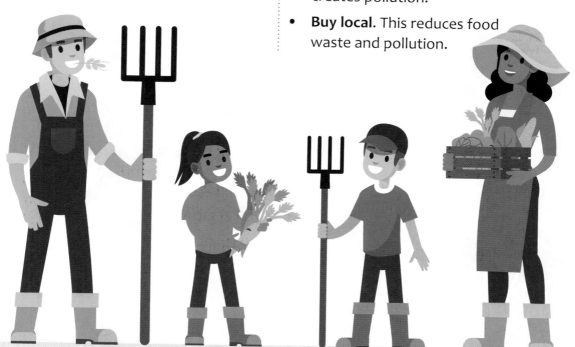

FAST FACTS

Since 2017 in the United States, **94%** of food scraps end up in landfills.

As of 2017, food and yard waste make up about **28%** of a household's garbage.

Composting can lead to **LESS GARBAGE**.

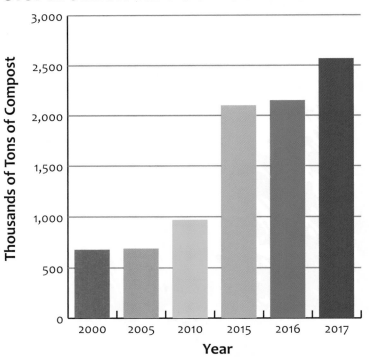

2019, U.S. Environmental Protection Agency

Activity

GROW A VEGGIE GARDEN

You can get involved in agriculture by growing your own vegetable garden.

1. Research what veggies grow well in your area. Lettuce, carrots, and radishes are usually good for beginners. You can pick out seeds from a local store or check to see if local gardening groups have seeds for free.
2. Write out a plan for your garden. When should you plant your seeds? Work with a parent or guardian to set up an area outside or a planter inside your home.
3. Plant your garden and keep track of its progress. If you have a problem with bugs, research natural ways to solve it.
4. Enjoy the fruits of your labor!

Learn More

BOOKS

Johnson, Robin. *Food Scientists in Action*. New York, NY: Crabtree Publishing, 2019.

Labrecque, Ellen. *Permaculture*. Ann Arbor, MI: Cherry Lake Publishing, 2018.

Smundak, Katharina. *The Truth Behind GMOs*. New York, NY: Rosen Central, 2018.

WEBSITES

Action for Healthy Kids
https://www.actionforhealthykids.org/activity/food-waste/

Britannica Kids
https://kids.britannica.com/kids/article/agriculture/352715

NeoK12 Agriculture
https://www.neok12.com/Agriculture.htm

BIBLIOGRAPHY

Climate Change Connection. "Agriculture Solutions." Last modified December 29, 2015. http://climatechangeconnection.org/solutions/agriculture-solutions

Food and Agriculture Organization of the United Nations. "Food Loss and Food Waste." http://www.fao.org/food-loss-and-food-waste/en

National Geographic. "Agriculture." Last modified October 9, 2012. https://www.nationalgeographic.org/encyclopedia/agriculture

The World Bank. "Agricultural Land." https://data.worldbank.org/indicator/ag.lnd.agri.k2

United States Environmental Protection Agency. "Greenhouse Gas Emissions." Last modified August 8, 2019. https://www.epa.gov/ghgemissions

GLOSSARY

composting (KOM-pohst-ing) using decaying material, such as leaves or vegetable scraps, to improve garden soil

deforestation (di-for-eh-STAY-shuhn) the removal of all the trees in an area

genes (JEENZ) parts of cells that carry information on traits such as appearance and growth

industrialize (in-duhss-tree-uh-lyz) to use factories and machines to produce what was once made by hand

irrigation (ihr-uh-GAY-shuhn) to supply crops with water by using pipes or channels

municipal (myoo-NISS-uh-pul) having to do with the government of a city or town

species (SPEE-seez) a group of animals or plants that share similar traits

yield (YEELD) how much a particular farm, field, or plant produces

INDEX

air, 14, 18, 19, 24, 25

composting, 26, 28, 29

emissions, 24, 25, 27

fertilizers, 19, 21, 27, 29

genetically modified organisms (GMOs), 7

herbicides, 14

industrialization, 7
irrigation, 7

land, 6, 8, 10, 14, 17

pesticides, 14, 18, 19, 20, 21, 22, 27, 29
pollution, 4, 14, 18, 19, 24, 27
population, 6

soil, 4, 21, 25, 29

waste, 14, 15, 16, 26, 27, 28
water, 14, 18, 19, 20, 29

ABOUT THE AUTHOR

Renae Gilles is an author, editor, and ecologist from the Pacific Northwest. She has a bachelor's degree in humanities from Evergreen State College and a master's in biology from Eastern Washington University. Renae and her husband live in Washington with their two daughters, Edith and Louisa.